M000288345

THE MATING GAME

Alex Cooper

Illustrated by Ingvild Hamre Konglevoll

"If you watch animals objectively for any length of time, you're driven to the conclusion that their main aim in life is to pass on their genes to the next generation."

Sir David Attenborough

"Could you give us a bit of privacy, David?"

Animals

After Alex's last book, things between us have been a little tense. (He ghosted my daughter Gertrude after three dates. She was inconsolable.)

As such, I was slightly surprised when he reached out asking me to write *another* foreword.

I was relieved when he told me there wouldn't be any more hypothetical animal fights in this book. That relief quickly turned to confusion, then to concern, when I heard what he was planning this time.

As a happily married silverback with a harem of my own, I have no need for a dating profile. (Although my sixth wife just got dragged off by a leopard, so there's a vacancy...) And I have concerns that most animals on the following pages are lacking opposable thumbs. Or, in many cases, *thumbs*.

In other words, I think Alex's animal dating app is a terrible idea, and I'm only doing him a favour because he caught me in an uncompromising position with my bonobo secretary.

I must admit, I was grudgingly impressed by the research that went into this project. Don't get me wrong, I was mostly concerned. And appalled. And even slightly aroused, in places.

After finishing *The Mating Game*, I've become a certified expert on my fellow creatures' love lives. Is that good or bad? I'm not sure. But I want you all to read it, so I have someone to discuss it with at my next group therapy session.

What I do know is that "penis fencing" has been added to my lexicon, otters and penguins have been absolutely ruined for me, and it's all Alex's fault.

I'm worried about you, buddy. Can't you just write a children's book next time?

Gary Gorillason

Dating profiles provide interesting insights into the human psyche.

For instance, a topless mirror selfie and a picture hugging a heavily sedated tiger suggest the person in question is a douchenozzle.

Meanwhile, someone whose only interests are "eating and hanging out with friends" likely possesses a personality marginally less interesting than a soggy Weetabix.

Amateur psychology lesson aside, it's easy to think we're the only ones who struggle in the game of love. But that's far from accurate. You may cry into your ice cream because Chad stood you up to attend the Tiddlywinks World Championships, or curse Stacey's name since she declined to see your Pokémon card collection, but you'll get over it. There's always next time.

When it comes to other species, it can—quite literally—be a matter of life and death. Animals don't have the luxury of focusing on their career or finding themselves during a meditation retreat in Ulaanbaatar. Their lives are predicated on passing on their genes, and there might not *be* a next time if they fail.

With the stakes this high, it's unsurprising that animals have evolved a bevy of tactics to win the mating game, which put even the most dedicated human dating addicts to shame. From perfect plumage to sneaky strategies, awe-inspiring artwork, and eyebrow-raising anatomical adaptations, these lovelorn critters will do whatever it takes.

Given the sheer wackiness of many animals' love lives, I thought it would be fun to see what would happen if they could write their own dating profiles. But they never responded to my emails, so I was forced to take a few creative liberties and write them myself.

Each animal in the book has its own illustrated dating profile, followed by a bit of background info explaining its mating strategies in more detail and/or apologising for all the terrible puns. Some facts are heart-warming; some facts will turn you into a shell of your former self, wandering deserted beaches and staring imploringly into the distance, searching for a sign that helps you make sense of a world in which four-pronged penises exist.

I hope you enjoy the book as much as I enjoyed writing it. But be warned—you probably won't look at your favourite animals in quite the same way afterwards.

Lion

Leo, 7

Freaky feline seeks self-sufficient spouse(s)

Ladies, this is what an alpha male looks like. You'll have to be okay with sharing this polygamous predator, but I need to be your mane man, and I've got no time for cheetahs.

My interests include sleeping, eating, sleeping some more, then showing you a good time for 12 seconds.

I want a woman who can take care of herself *and* take care of dinner. I'm happy for you to spend some time with the girls...as long as you're out hunting.

And one more thing—I'm not raising another guy's cubs. (No, seriously. I will kill and eat them in front of you.)

Favourite Quote: *"A lion doesn't concern himself with the opinions of sheep"* — Me

L eo's casual sexism may seem like something straight out of a 1960's ad agency, but equality of the sexes is a foreign concept for much of the animal kingdom.

For human males, brute strength and aggression are no longer essential when it comes to finding a mate. Women are just as likely to go for the sensitive Fabrizio—who writes poetry and plays three chords on the acoustic guitar—as they are Big Dave, who has three teeth and gets in fistfights at the pub every weekend. (Or perhaps *more* likely.)

For male lions, dominance is crucial. Young males grow up in a pride, nursed by various females. Before long, they're forced out to fend for themselves and find new territory. If they can't, they'll live as nomadic males without a place to call home. (If they had the manual dexterity, maybe *they* would attempt some sad acoustic songs.)

Many lions team up, forming coalitions with other males, usually family members. They've got a better chance of displacing rivals or defending their pride with a numbers advantage.

At the age of just four or five, a lion could be overthrowing an incumbent male, while his human counterparts are eating crayons. By the time Timmy has his first wet dream, a male lion's life will be winding down.

The intervening years will be tough, even for the small percentage of males who win and retain territory. It's true that male lions sleep a lot, but they need to keep their strength up for patrolling their domain and fending off rivals. The lionesses will do the lion's share of the hunting, but males sometimes help with bigger game, where their larger size is helpful.

And the stuff about lions being bad stepfathers? Absolutely true. If challengers successfully defeat the reigning resident males, they'll kill any cubs they find.

It's not because Leo is a monster; it's simply his biological imperative. While the lionesses are nursing his potential future rivals, they won't be receptive to mating. By killing the cubs, he's killing two birds with one stone.

He can then set about making some babies of his own with the recently bereaved lionesses. Talk about a mood setter.

As for Leo's chances of finding a mate? Even with his slightly sexist world view, they look pretty good. He has a dark mane, which indicates high testosterone and a better chance of success in fights against other males. (Maybe Scar should sue Disney for defamation.)

Female lions are drawn to these darker manes, much like Fabrizio's perfectly groomed beard draws admirers as he plays "Wonderwall" at the open mic night. *Unlike* Fabrizio's beard, Leo's mane is not just for show. It also protects his neck in a fight, which—unless his guitar playing is exceptionally bad, and Big Dave is in the audience—Fabrizio won't have to worry about.

Angela, 12

Why are all men such parasites?

There are plenty of fish in the sea, so I'm casting out my lure and looking for a real man who won't leech off me and my successful fishing business.

I'm a big girl, so you must be at least 50 mm tall. You can't call me shallow—I live at the bottom of the ocean.

Guys always get attached too soon. I'm flattered by the attention, but I want someone who has his own goals (and his own gonads). And I'm sick of men only wanting me for my body. (I do believe that marriage is a fusion of two fish, but only metaphorically.)

In my spare time, I like floating aimlessly in the abyss and pretending to be a lamp.

Biggest Turn-Off: Being overly dramatic. You won't die if you can't be with me.

S exual dimorphism is common in nature, but anglerfish take it to the extreme.

Females of many species only reach a few centimetres in size, but they still dwarf the males. So much so, that scientists were confused when they kept catching female specimens with small parasites attached.

Those "parasites" turned out to be male anglerfish. So, next time you complain about a clingy partner, consider yourself overshadowed.

Angela should be successful at luring in both males *and* her dinner, thanks to the luminescent organ, called an esca, that's located on the end of her "fishing rod" (a modified dorsal ray). Tiny bacteria live on the esca, and they're what give it its creepy glow. Males don't have an esca, but they can be drawn to the light show. Some have large eyes, while others sport sensitive olfactory organs that detect a lady's scent.

Because anglerfish encounters are rare, there aren't many mating opportunities. The tiny males float about looking for females and will die if they don't find one. (So maybe Angela could cut them some slack.)

Males of some species lack fully functioning jaws and can't catch prey. They don't even have fully developed gonads, and they'll only mature if they attach themselves to a female. (I'm sure plenty of women are looking at their partner and nodding in agreement.)

When I say the word "attach," I don't mean it metaphorically.

The minuscule male—having found a female—chomps down on her skin and releases enzymes that basically fuse his face into her body. After that, he's completely dependent on his female host, sharing her circulatory system and benefiting from the nutrients she gains from feeding. All he offers in return is the ability to mate. (Alright ladies, you can stop nodding now.)

Interestingly, some females can collect as many as eight of these little sperm donor freeloaders. Other species limit themselves to a single fun-sized soulmate, while yet more anglerfish don't fuse in this way at all. But that's no fun...

I think the love life of the anglerfish is almost touching, albeit in an "if M. Night Shyamalan did rom-coms" kind of way.

Orangutan

Orville, 18

Oobee doo, I wanna be with you.

Leave those lesser apes and take a trip to my treehouse of love. You can even share my orangu-tanning bed. (How else would I maintain this glorious ginger tinge?)

I'm a solitary soul, but even arboreal apes like me need company sometimes. If you're a frisky frugivore, just give me a call.

Actually, let *me* call; I've got the throat sac for it.

As the resident male in this forest, I like to display my dominance. That means chasing off transient troublemakers and flexing on them with my flange.

Interests: I heard we're endangered. I'm keen to start a little conservation project of our own.

F lange.

Flangey flangey flange.

What's that? Ah, no. I'm not going crazy; I'm merely admiring Orville's magnificent cheek pads, otherwise known as flanges.

It may not look like much to hot-blooded humans, but let me tell you, female orangutans go crazy for a cheeky chappy.

Why? Good question.

Flanges are still considered a bit of a mystery, even by orangutan experts. This facial feature isn't ubiquitous in males, and there's no consistent pattern to its emergence.

One day, a sub-adult Orville is chilling in the canopy, looking like an adolescent. The next, he feels *different*. A change is coming.

He feels it in his...*face*?

That's right. Orville suddenly starts to sprout a set of cheek pads made of fatty tissue, which make his orange rivals green with envy.

It is as the prophecy foretold. He has become a satellite dish.

To attempt to understand the flange, we must first understand the creature upon whom it grows. Let's go back to the beginning.

Orangutans are more solitary than the other great apes, although they can be somewhat social.

Like many species, the child-rearing duties fall to the female, and the male plays no part. It's a tough gig; childhood dependence for orangutans is one of the longest in the animal kingdom. There's a lot to teach them, after all.*

Babies will be carried everywhere for several months. Slowly, they'll learn to climb for themselves and begin to socialise. Nursing lasts up to eight years, and adolescents often stay nearby for a few years afterwards.

Flanged males are the most solitary of all orangutans. Their interactions are mostly limited to brief flings with females or aggressive encounters with other males.

Adult orangutans *do* essentially have a treehouse. They make new nests in the canopy every night, with larger branches as the foundation and smaller branches forming a mattress of sorts. (So, technically it *is* an orangu-tanning bed, but Orville's colouration is all natural, baby.)

Orville's impressive throat sac lets him bellow out booming calls that carry long distances through the canopy. This can serve as both a warning to other males and a signal to receptive females.

Unflanged males will likely avoid the sheer manliness of a physical specimen like Orville. If another flanged male is in the area, though, they could come to blows over mating rights.

So, *why* do some males start to grow flanges?

It's clear that it gives males a sexual advantage. Orville is an orangutan's orangutan. Guys want to be him; girls want to be with him. It's not simply a sign of sexual maturity, though. Males can mate many years before the first flourishing of flange. (Although they're less likely to impress discerning ladies.)

The flange is just one part of a significant physical transformation. As well as this propitious increase in facial surface area, he'll also get his throat sac for vocalisations, start growing longer hair, and develop an irresistible musky scent.

What causes him to become this arboreal alpha male, though?

A study of wild Bornean orangutans found increased testosterone levels in flanged males compared to their unflanged counterparts, though the highest levels were found in a developing male during flangification (if it isn't a word, it should be). So, it's clear that high testosterone levels are needed to make the transformation.

The trigger for this testosterone surge is more of a mystery. Why do some males strike the flange jackpot, while others are stuck in arrested development? Experts aren't entirely sure, but the mere presence of a flanged male can suppress flange development in other males.

With Orville ensconced as his area's resident badass, the prospects aren't so good for unflanged rivals. They can wait for him to pop his clogs, or they can become transient males and move to a new territory. If all goes well—and they're not emasculated by another flanged male—they'll probably start to grow their own.

In conclusion, I have just one more thing to say:

Flange.

Adele, 6
Waddle it be?

I need a passionate pebble provider who'll take care of me on Adélie basis.

My ex was too much of a gentoo man. He tried to make friends with a leopard seal, leaving me to raise a chick on my own. (If the thought of babies makes you puke, I'd love to hear from you. She's hungry.)

Trust is important in a relationship, so stop questioning my motives. I'm not just looking for a feathery fling; I want a rock. *A pile of them.*

That doesn't make me a pebble digga, but I do like a man with status. If any Kings or Emperors are reading...hit me up.

Favourite Movie: *Happy Feet.*

A h, penguins. Few animals are as wholesome as these wacky waddlers.

What I hope you've learned by now is that even the silliest, most adorable animals have their character flaws. Penguins are no exception and can lie, steal, and cheat with the best of 'em.

"But I read about penguin pebble proposals," I hear you say. "Isn't it an avian romance for the ages?"

Well, there *is* an internet factoid that the male penguin will search for the perfect pebble to present—like an engagement ring—to his mate. This isn't *exactly* true, but it has some basis in fact.

Several penguin species build nests out of small stones. Both males and females will search for pebbles and use them to construct the nest. So, Adele the Adélie wouldn't turn her beak up at a rock for being too small, and her suitor wouldn't comb the beach with his jeweller's loupe at the ready.

Pebble presentations are still a big deal, though, and can be accompanied by head bowing, vocalisations, and other bonding behaviours during courtship.

Adele might look down on a potential mate for not having *enough* pebbles. Adélies need an elevated nest to protect the chicks from drowning if there's rain or if the snow melts. So, it's a case of quantity over quality.

It's not being choosy for the sake of it; she needs a provider who gives her young the best chance of survival. (That's also where the puking part comes in—adults regurgitate a delicious pre-digested fishy treat for their chicks.)

Adélie penguins have a loose grasp of personal property rights. A male will covet his neighbour's nest and pilfer a pebble or two if it gives him a better shot at breeding.

Some females also have loose morals and will even turn to prostitution. A BBC article from 1998 recounted how female penguins would cheat on their partners by trading sex for stones, so perhaps Adele isn't as innocent as she seems.

Gentoo penguins, as well as being prime pun material, are a related species. They too have a predilection for pebbles, which they need for their nests. I don't *think* they try to make friends with leopard seals, which are notable penguin predators.

Just be thankful the dating profile wasn't that of a young male Adélie. Their "sexual depravity," documented during Captain Scott's Antarctic expedition, was considered too risqué for publication at the time and was only unearthed decades later.

George Murray Levick was the scientist who spent a memorable summer in 1911–12 watching these penguin perverts, who would attempt to mate with...well, *anything*, really. Some males would try their luck with other males, young chicks, or even dead females.

Levick wrote a paper about the penguins once he got back to England, but the naughty parts were considered too lewd and were left out "to preserve decency."

"There seems to be no crime too low for these penguins," Levick wrote, after (probably) taking a sniff of his smelling salts.

Maybe Adele should set her sights on a more mature gent(oo).

Penelope, 10
What's the rush?

I'm an ex-carnivore, but since I moved to a vegan diet I feel so much better and more energe... sorry, just dozed off there.

You'll have to deal with my daddy issues and identity crisis. My father was an anonymous sperm donor, and it's tough being Black, White, *and* Asian. I got a great modelling gig for the WWF, but I wonder whether they're just pandering because I'm cute and exotic.

I've met too many bamboo guzzlers who were bamboozled by the sight of a woman. Let's let things grow organically, like our dinner.

Date Idea: Maybe we can watch a naughty movie to get us in the mood. (At the end of our 100th date.)

G iant pandas are members of the bear family (Ursidae), but they've branched out on their own when it comes to diet, despite remaining in the order Carnivora. Their fearsome ancestors would probably shake their heads at what's become of their lineage.

The panda's diet is 99% bamboo, with only the occasional animal-based "cheat meal." The trouble is, its digestive system is still that of a carnivore, meaning it derives little energy or nutrients from its raw, plant-based diet. This contributes to the panda's slow metabolism, largely sedentary lifestyle, and inability to write a dating profile without dozing off.

Wild pandas are mostly solitary creatures, but will get together for the mating season, between March and May. They'll give birth to either one or two cubs, but if it's twins, the weaker sibling usually won't make it. The mother simply doesn't have the resources to raise two children.

With their scarcity in the wild, much of their breeding is done behind bars. (Zoos, not prison.) The trouble is, captive pandas don't show much interest in the opposite sex, and getting these lazy layabouts in the mood is easier said than done.

This forced Chinese scientists to get creative. Frustrated with the charismatic critter's lacking libido, they started showing "panda porn" to captive animals in the hopes they would pick up a few pointers.

And what would panda porn be without some "performance enhancers"? Yep, you can't make this stuff up—they gave male pandas Viagra.

The strategy got off to an inauspicious start, with 16-year-old male Zhuang Zhuang the first to receive the experimental treatment, which had "no result on him at all."

"We used the wrong panda. That panda basically has no capability," said Guo Feng, scientific researcher and destroyer of ursid egos.

Despite the best efforts of mean scientists, most panda breeding is done via artificial insemination. That explains Penelope's (lack of) relationship with her dad, who probably came from a test tube.

Giant pandas are essentially the definition of "charismatic megafauna," as a large, iconic endangered species with widespread conservation appeal. Perhaps Penelope is a little insecure about whether she's deserving of all the money and resources allocated to panda conservation, when so many smaller, less adorable animals are also struggling. (No one's making an endangered toad the logo for the WWF...)

Let's hope she finds a lover who appreciates her for what's on the inside.

A whole lot of bamboo, in varying states of digestion.

Koala

Kayleigh, 5
Are you koalified to handle me?

I'm not like other girls; I'm a go-getter who only sleeps 18 hours a day. #Onthegrind

I want a man with brains *and* brawn. You have to fight for me, but I've had more engaging conversations with a eucalyptus leaf than my last few dates.

I go weak at the knees for a deep voice and a potent stench. That's how I fell for my ex-con ex-boyfriend, who was a member of the Gum Tree Gang. (He said he was innocent, but his fingerprints were all over the crime scene.)

Oh, and if things get serious, don't be offended if I ask you to get tested for STDs. I don't want to survive one kind of bushfire just to get another.

Cats or Dogs? I hate both.

I t seems we've gone from one lethargic lover to another. Maybe Kayleigh is the exception. "Go-getter" and "sleeps 18 hours a day" don't usually feature in the same sentence, but compared to her peers she's a real workaholic.

Koalas are largely sedentary and can sleep up to 20 hours a day; their diet doesn't allow for much else. Eucalyptus is so lacking in nutrients and calories that a few hours of lazy grazing is all most koalas can manage.

It's not much fun to eat, either. The tough leaves grind down ko-alas' teeth, which don't regrow, causing older individuals to starve to death. (How are these animals not extinct, again?)

Kayleigh's desire for an intelligent companion is commendable, but she might be setting her sights a little high. Koalas have one of the smallest brains (proportionally) of any mammal, at just over 19 g on average.

Their brains are also remarkably smooth, which may sound nice, but those trademark wrinkles in the brains of humans (and many other mammals) allow for a larger surface area in relation to vol-ume. In more technical terms, that's the fing wot makes us smart, innit.

If you plucked several delicious eucalyptus leaves and put them on a plate in front of a koala, it wouldn't even recognise them as food. Koalas put two and two together and get...exhausted. Then go to sleep hungry.

Finding someone who'll fight for her is a more realistic goal. Male koalas will tussle over breeding rights, and older males have the scarred mugs to prove it. Many don't reach old age—getting flung from a tree is quite hazardous to one's health.

Kayleigh likes a masculine marsupial, and it's understandable. If you ever hear a male koala's bellowing mating call, you'll be surprised at how loud and low-pitched it is. (And possibly a little turned on. I don't judge.)

A male's chest gland is also used to rub against a tree and mark his territory. I haven't smelled these secretions myself, but female koalas have told me they go crazy for it.

Our eligible young Aussie seems to have sworn off bad boys, but maybe her ex *was* framed. Koala fingerprints are so similar to those of humans that experts would (supposedly) struggle to tell them apart under a microscope. (Even if it's true, I'm not sure it's a pressing issue for forensics departments.)

I suppose it depends on the crime—if it were a eucalyptus heist then the Gum Tree Gang would be the prime suspects. If officers investigated a spate of armed train robberies, then Bruce Cassidy and the Sunshine Coast Kid could probably be ruled out.

As for her sexual health? She's right to be cautious, because many koala populations are rife with diseases like Chlamydia. (Lewd innuendo aside, bushfires have sadly decimated koala numbers. As have feral cats and dogs, which could explain Kayleigh's disdain for them.)

After Kayleigh and her new beau have had a fumble down under, a tiny joey will be born. It immediately crawls into her pouch, where it will live for the first six or seven months of its life.

Gross fact: *As joeys move from their mother's milk to a eucalyptus diet, there's a transition period where they eat a "faecal pap" of digested leaves from their mother's cloaca.*

By the time Kayleigh's little bundle of joy reaches its first birthday, it will be fully weaned and spending more and more time away from her. That's good, because if she gets pregnant again, she'll aggressively encourage her adolescent child to take a hike and find its own territory.

It's a good job they're cute.

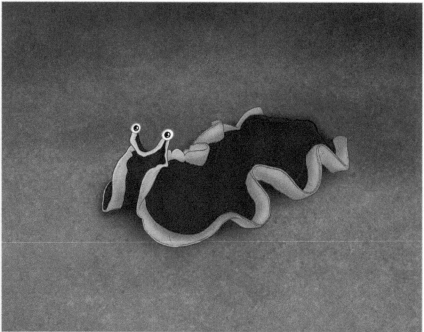

Fran, 2 months
En garde!

This aquatic fencing fanatic is looking for a duelling partner who's willing to risk it all. The winner gets the spoils; the loser gets the sperm.

Let's not beat around the bush. We're both horny hermaphrodites, but one of us has to give birth. We *could* flip a coin, but where's the fun in that? Let's wave our willies around until one of them hits the target.

My last relationship was quite traumatic—my *flat mate* knocked me up and left me. I'm sick of being a mother, so don't expect me to be chivalrous and let you win.

Celebrity Crush: Michael Flatworm, star of *Oceandance*. His footwork is superb.

Y ou're probably fairly confused. I guess the term "penis fencing" is as good a place to start as any.

Many species of flatworm (and all that engage in this spermy swordplay) are hermaphroditic, having both ovaries and testes.

Some species will transfer sperm perfectly amicably and mutually inseminate one another, known as bilateral sperm transfer.

Others, like our friend Fran (who is a *Pseudobiceros hancockanus**), prefer to liven things up with a spot of healthy competition. Only the loser will become the mother; the winner, or "father" can swim away without a care in the world. (Child support hasn't caught on in invertebrate society.) This is called unilateral sperm transfer.

*Hancockanus. Hancock…anus. What was that guy doing with the flat-worms?

When it's time to mate, the flatworms whip out their extendable, go-go gadget "penises," which are actually sharp, dagger-like structures called stylets. They then attempt to pierce their mate's skin and inject them with sperm from either of the stylet's two heads.

This is an example of "traumatic insemination," so poor Fran was being quite literal when talking about her* last relationship. Getting a hole gouged in you by your lover's nether parts is about as much fun as it sounds.

*Or "his" —if he wins his next fight.

31

I'm not sure how the fencing starts, and it probably doesn't involve a French referee asking whether they're ready, but the two opponents will spend up to an hour looking for the opportune moment to knock up their rival.

Of course, each flatworm is trying their best *not* to be inseminated. Quite apart from the physical damage the act can cause, being a mother requires much more investment in terms of time and energy. Every cloud has a silver lining, though—if the loser's offspring take after their daddy, they could be excellent duellists.

The lure of competition is too strong for some fathers. Much like an undefeated fighter who can't bring himself to go out on top and retire, undefeated flatworms will often continue to penis fence until they're impregnated.

Live by the penis sword, give life by the penis sword. That's in the Holy Flatworm Bible. Probably.

Petunia, 6 months

We've all got a few exoskeletons in the closet.

Father forgive me, for I have sinned. Yet I keep reinstalling this app.

I'm not the young nymph I once was; I'm at the stage of life where I need to spread my wings and find a man who's worthy. I don't want a ladykiller; I'm already a *lady killer* and kung fu master. It's time I traded martial arts for marital bliss.

This penitent predator is partial to a partner who won't lose his head at the prospect of getting serious...but maybe I have my own commitment issues to work on.

Favourite Song: "Where's Your Head At" by Basement Jaxx, or anything by the Fine Young Cannibals.

I took the low-hanging fruit there with all the praying jokes, but people sometimes get confused and call them *preying* mantises. Understandable, since they *are* voracious predators, but mantises get their (common) name because of the way they hold their forelegs in front of them. (I've never heard of a mantis going to confession. If they *are* remorseful, they do a good job of hiding it.)

I think many of us know the story about female mantises eating the male after mating, but it's not something that happens every time in the wild. It's hard to tell how often it occurs, but it's unlikely that more than a quarter of encounters end in the male's demise.

I still don't like those odds.

In captivity, for whatever reason, sexual cannibalism is a lot more common—even more so when the "hangry" female hasn't eaten recently. In certain cases, with a well-fed female and an absence of scientists peering at them, the insect lovers engage in elaborate courtship rituals. Sadly (for the males) most captive mantis couplings are more horror story than love story.

Mantises love to start consuming their prey head first, and peckish mating females are no exception. The male, to his credit, soldiers on gamely after losing his head and will continue to mate despite this small setback. In fact, the male's movements may even become "more vigorous."

Now I see why the females do it...

Petunia would make a great martial arts instructor. Praying Mantis Kung-Fu is a well-known style based on the tactics and movements of the insect. (I don't think it includes beheading, though.) Mantises, with their rapid reflexes, are the perfect ambush hunters; many an insect has had its life snuffed out by those serrated forelimbs.

Petunia is also, quite literally, at the right stage in life for dating.

Mantises go through egg, nymph, and adult stages. Petunia moulted several times on the way to her final form; only now has her mind started wandering to affairs of the heart. (And the head...*the eating of it*.)

The whole process can take several weeks to several months, depending on the size of the species. Larger species like Petunia, a European mantis, take longer to mature.

When it's time to mate, a male will approach slowly and carefully, which seems eminently reasonable. The female is significantly larger, and he doesn't know when she last ate. He'll creep up slowly from behind, sometimes taking hours to reach his lethal lady friend.

If he's lucky, he'll jump on her back, do the deed, and hop off again before she has time to decide he can stay for dinner.

If not? Well, the male will usually mate for *longer* after being beheaded. His body can survive for a few hours, and he's less inclined to climb off and call a cab once his brain has been eaten. (The female also seems perfectly happy to let him go at it for a while.)

If it's any consolation, his body will provide her with nutrients to create a bigger ootheca (egg case) with larger eggs. Let's hope Petunia can find a selfless dad like that.

What, no volunteers?

In the immortal words of my favourite insect poet, Backyard Kipling:

"If you can lose your head when all about you are keeping theirs...you'll be a mantis, my son."

Sam, 2

Waiting for Cupid's love dart to penetrate me.

I'm shy at first, but after a few dates I start to come out of my shell. And I'm going to be straight with you—I'm gender fluid. But then again, you are too.

Love hurts, but I'm willing to take a stab at it again. Foreplay is important, though; let's feel each other out and take our time with the slime.

As a successful gastropod gastro pub owner, I need someone who's got their life together. That means a mobile homeowner. (Sorry slugs.)

And NO LOVE DART PICS! I'm not that kind of mollusc.

Favourite Song: "Escargot Your Own Way" by Slowwood Mac.

Despite occurring in your garden, this is no garden-variety romance. It's the sequel you neither asked for, nor wanted: *Punctured by Passion: Return of the Love Stabbings.* Once again, we've got a hermaphroditic invertebrate with a saucy sex life. The star? The humble garden snail, or *Cornu aspersum*. I don't want to cast aspersions on the *aspersum*'s character, but this gruesome gastropod has some explaining to do.

Flatworms may fence with their sharp schlongs, but snail courtship is a similarly risky endeavour that often ends up with someone getting stabbed. It's not a snail's *penis* doing the stabbing, though; it's their charmingly named "love dart."

If you're imagining a giggling gastropod Cupid gleefully firing at slimy sweethearts, then I want some of what you're taking. But also, you couldn't be more wrong, for it is clear that romance has forsaken snail seduction.

The love dart is a sharp, calcium carbonate (or chitin) structure that some snails and slugs produce, which is stored in the aptly named dart sac. Unlike the flatworm's stylet, it isn't used for insemination, but it does play a role in reproduction.

Love darts come in a variety of shapes and sizes, but they're all sharp and covered in mucus. (It just gets better, doesn't it?) Inside that mucus are hormones that reconfigure the "female" snail's reproductive system and make her more receptive to the "male's" sperm.

The courtship ritual progresses at a snail's pace. The main event is preceded by several hours of tactile foreplay, which involves tentative tentacle touching and lip biting. The sneaky snails are just feeling each other out, waiting for the right moment.

Meanwhile, the pressure is building. *Literally.*

As the snails circle each other lovingly/warily, hydraulic pressure builds around the organ housing their love darts. They manoeuvre themselves into position, like Wild West slime slingers looking to get the first shot off.

Then it happens. One snail's body touches the other's genital pore, and this triggers the shooting. Usually. (Garden snails aren't *always* armed.)

The species that grow love darts don't start producing them until after they've mated once, so unarmed virgin snails are unlikely to enjoy their first time. Unless, that is, their partner already used up their own dart on a previous conquest. (Garden snails take a week to regrow love darts.)

Snail mating is basically Russian roulette—you never know what your partner's packing.

Even if both snails have their love darts at the ready, there's no guarantee they'll hit the target. Garden snails are especially inaccurate and can fail to penetrate skin—or miss their lover completely— as much as one third of the time.

Maybe it adds to the excitement.

On second thoughts, probably not. Sometimes a love dart will be fired so forcefully that it lodges in the snail's internal organs or goes right through its body like an escargot kebab.

Once the snails have got their shots off, they'll proceed with mating. As hermaphrodites, the snails can fertilise each other's eggs, although sometimes only one snail will transfer sperm. (Occasionally, garden snails dispense with mating completely and self-fertilise their eggs, which makes you wonder why they run the love dart gauntlet in the first place.)

Being run through with a mucus-covered projectile has its downsides. (Shocker, right?) In what seems like a counter-intuitive move for the survival of the species, the traumatic love stabbing shortens the lifespan of the recipient and reduces her fertility for future matings.

So *why* try and shoot each other, especially when the snails could peacefully transfer sperm?

If you dig deeper, it sort of makes sense. The hermaphroditic snail's selfish male half wants to ensure it's *his* sperm that's being used to make the next generation of slimy stabbers. The hormones released with the darting help him out by causing more of said sperm to fertilise his partner's eggs.

It seems that these land snails are willing (or evolutionarily programmed) to risk a shorter life, because the benefits of being quicker on the draw (more offspring) outweigh the consequences of getting stabbed (pain and misery).

Scientists predict that this sexually antagonistic arms race will lead to a "violent escalation of sexual conflict," as the snail's male and female parts wage war with one another.

I think Sam needs to see a shrink (a miniature psychiatrist) about those self-destructive tendencies.

Eric, 28
Everyone else is irrelephant.

The dry season is over, ladies, because this young bull is on the pull.

I used to be polite and respectful, but then something snapped. You may say you like nice guys, but sooner or later you'll fall for this hunky piece of trunk. (And my trunk's not the only thing that's prehensile, if you catch my drift.)

When I'm not working on my tree felling business or crashing rhino bars and starting trouble, I like looking in the mirror and being self-aware of how awesome I am.

They say you are what you eat—I guess that's why they call me the African *bush* elephant.

Favourite Chat Up Line: What's long, wrinkly, and dangles between my legs? Not my trunk.

I know what you're thinking: Eric's a bit of a douchebag.
Well, it's not entirely his fault. Like all sexually mature
bull elephants, he sometimes suffers from a condition called
musth, where sex hormones course through his blood, and he becomes irritable and unpredictable.

During musth, a bull elephant's testosterone can increase to over
100 times normal levels. An unpleasant, sticky substance called
temporin seeps down the sides of his head and drips into his mouth,
while his temporal glands swell up, putting pressure on his eyes and
causing acute pain.

No wonder he's moody.

Many males start to undergo this process at around 25–30 years
old. Younger bulls will have shorter periods of musth, lasting days
or weeks, but as they age it can go on for several months. Eric's still
a young bull, so his braggadocio could be a cover-up for his confusion over this Jekyll and Hyde transformation.

Male elephants in musth are often responsible for attacks on
other animals, including humans. Older males mellow a bit and
keep the younger bulls in check, but horny young bachelor herds
have been known to run riot, leaving trampled trees and crops, vil-
lages, and dead rhinos in their wake.

The mere presence of older, dominant males can suppress musth
in younger bulls. National parks in South Africa have had success
introducing older males to bachelor herds of misbehaving adolescents, much to the delight of all rhinos in the vicinity.

Perhaps Eric is acting out because he doesn't have a father figure
to look up to.

If you *didn't* catch Eric's drift in his profile, I don't blame you, because he was talking about his prehensile penis. (Yes, my internet search history *is* a mess.) Male elephants exhibit an impressive amount of manual dicksterity (not a typo) with their "fifth leg," even using it as a weight-bearing aid to prop themselves up or scratching their belly with it.

It has other benefits. For a several-tonne animal, mating isn't exactly easy (or comfortable). Having more control of his substantial schlong helps Eric direct it where it needs to go before he inconveniences his mate too much.

That's not the only unusual thing about the massive mammal's monstrous member.

During musth, elephants can get something called "green penis syndrome." And no, it's not something Bruce Banner struggles with when he transforms into the Hulk; it's discolouration caused by a steady stream of urine that drips down his manhood.

Males will strut around, showing off their dribbling green peen to all and sundry in the savannah, and hoping the ladies like what they see. Weirdly enough, *they do.* Males in musth have more success attracting elephant cows in heat.

Females don't just fall for the first gesticulating green penis they see, though. They like an elephant who's going grey. Err, more grey than usual, that is. *They like older men,* is what I'm getting at. Given the choice of a young bad boy like Eric or a distinguished silver fox, they'll probably opt for the bigger, older, and more dominant bull.

As for the mirror stuff? The mirror test is used to gauge animal intelligence and indicate their level of self-awareness. (Most species don't realise it's their reflection.) Tests have been performed on Asian elephants, and while there were mixed results, one individual repeatedly touched an X on her head with her trunk—a mark which could only be seen by looking in the mirror.

So, we don't know whether Eric, an African elephant, is self-aware, or just extremely vain.

Perhaps both?

Hopefully, he'll grow out of his youthful exuberance and become a wise mentor to the next generation of trunky troublemakers. But for now, let's give him a wide berth and hope he learns some manners in the next few mating seasons.

Antonio, 1
Can we mate, mate?

I'm a fair dinkum dreamboat who's small in stature but big in stamina. My whole life has been leading up to this moment, and I've been saving all my love for you. (And your friend. And your sister. And your second cousin. And your netball team.)

This is a one-time offer. I'm here for a good time, not a long time, so you'd better pouch this macho marsupial before bits start falling off.

I won't be there to see Junior grow up, and that haunts me every day, but I know you'll be an awesome single mother.

I'm just a boy. Standing in front of a girl. Telling her to QUIT PLAYING COY AND GET ON WITH IT!

Favourite Movie: *The Expendables.*

Antechinus sounds like a tragic hero from Greek mythology, but it's actually the name of a group of mouse-like Australian marsupials.

I guess Antonio the Antechinus *is* the tragic hero of his own sexy, short-lived tale. Because he—quite literally—goes out with a bang.

The male Antechinus got screwed over by evolution, so he decided to screw right back. He only lives for one breeding season, and for a three-week window when the females are in estrus, he takes full advantage...only to die at the end.

He shows admirable stamina. Mating can be "intense" and "last up to 12 hours," according to a paper in the *Australian Journal of Zoology*. Antonio will be up night and day for the breeding period, trying desperately to pass on his genes to as many gals as possible, even as his body breaks down.

You see, he wasn't exaggerating (much) when he mentioned bits falling off.

His body ramps up production of free corticosteroids in his blood, allowing him to tap into extra energy reserves for this mating marathon. This comes at a cost—his immune system is suppressed, and he'll be wracked with ulcers, before finally shuffling off this mortal coil with a proud smile on his little face.*

*I could be anthropomorphising slightly. (Not for the first time.)

But *why* has this self-sacrificial mating strategy evolved?

It's thought that, even if mating *weren't* fatal, the males probably wouldn't survive until the next breeding season. So, it makes sense for them to take full advantage of this short window and put all their effort into spreading their seed far and wide.

Females, on the other hand, may live to see one or two more breeding seasons. They'll give birth to large litters which have multiple fathers, making it difficult to know which tiny headstone the widowed mothers should lay a wreath at.

During this period of miniature marsupial madness, Antonio will be blissfully free of the usual worries that preoccupy an Antechinus, and he'll become laser focused on mating with as many females as possible.

It seems strange, but it offers him the best chance of passing on his genes. It must work, too, because these naughty little scamps are still around.

"Survival of the fittest," in simplistic terms, refers to fitness as reproductive success. In this case, the "fittest" males are hopped up on hormones and on death's door.

I don't know whether to pity Antonio or envy him.

Steve, 2
Real men give birth.

Sick of those sleazy sea stallions who catch your eye then leave you for a guppy half your age? Fed up with the restrictive, patriarchal system that enforces rigid gender roles?

In that case...*how you doin*?

Parenting is hard, but it doesn't have to be. Why not let me share the load? Heck, just *give* me the load—I'm feeling broody. You can be a hands-off mother, but I'm not going to be an absent father. (Well, not until those little buggers pop out. Then they're on their own.)

I come from impeccable breeding. My father won the 3.20 at the New York Aquarium, in just over 26 hours.

Favourite Chat Up Line: How do you like your eggs in the morning? Deposited in *this guy*.

N o book on animal love lives would be complete without mentioning the seahorse. In a world of absent fathers, it's heart-warming to see a dad who's not only present at the birth of his children but is the one *giving birth*.

Let's back up a bit and see what leads to this unusual situation.

Seahorse courtship is long, weirdly romantic, and just plain weird. There's some pointing, and plenty of quivering, before the female finally deposits eggs into the male's brood pouch, which he leaves open for just six seconds. That's long enough for the sperm and the eggs to get acquainted in a "seawater milieu," which sounds like a trendy neighbourhood where fish work on their novels in soggy cafes, but isn't.

The doting father will then provide the fertilised eggs—now embedded in the pouch wall—with nutrients and hormones. (He produces prolactin, the same hormone responsible for milk production in mammal mothers.)

His lady love will come and visit him during the gestation period, to check in and/or gloat. While seahorses aren't strictly monogamous, many species will stay pair bonded for the breeding season.

The eggs are hatched in the brood pouch, and then the miniature seafoals* are released into the open ocean and left to fend for themselves.

*Not what they're actually called, but I think it has a nice ring to it.

The father will usually release several hundred young (depending on species, it can range from 5–2500), which are tiny, helpless, and prone to predation. Only a tiny fraction will survive, but it's okay, because he'll try and mate again soon afterwards.

Ah. Maybe not so heart-warming after all.

As the sharp-witted among you may have realised, seahorses are not related to *actual* horses. They're basically modified pipefish.

Ah, but what's a pipefish? It's, erm, basically an un-modified seahorse. (They're not related to *actual* pipes.)

I don't foresee seahorse racing becoming a spectator sport, despite Steve's boasts—they're notoriously poor swimmers, and spend a lot of time with their prehensile tail wrapped around some coral or seagrass while they have a breather.

Not to mention, it's a nightmare finding little shrimps to ride them, because seahorses tend to eat their would-be jockeys.

Beatrice, 3
Stop asking for booby pics.

This Galapagos gal loves to dance, so if you're a flat-footed featherbrain it's not going to work. Flashing some flamenco footwork or doing the sardine salsa always gets me in a flap.

I have trust issues. My ex was a wandering albatross with wandering eyes, and I found other boobies on his internet search history.

My dream is a clifftop condo with a couple of chicks. (Actually, maybe just one. Sibling rivalry is a nightmare.) If you're sitting on a nice little nest egg, then that's a bonus.

Favourite Band: Destiny's Chick.

Well, there's nothing at all funny about this noble bird. What are you tittering about?

Ah, who am I kidding? It's the simple things in life you cherish, and there are few things simpler than a bird with a silly name. But this is a serious, scientific tome, so I'll be on my breast behaviour from now on.

Sorry.

Let's address the elephant in the room. The bird's name supposedly comes from the Spanish *bobo*, meaning "stupid" or "clown," since those huge feet make them clumsy on land. The birds also used to land on ships filled with hungry sailors, which may have contributed to the nickname.

As you can see, Beatrice is sporting a couple of fantastic footsies with a vibrant blue hue. They're not just for show; foot colour is an important part of the mate selection process. That's because, as well as looking lovely, a bright blue foot indicates a healthy bird.

Their unusual foot colour comes from carotenoid pigments—obtained from the booby's seafood diet—which stimulate the bird's immune system. They've got to stay on top of things, because even a couple of days without adequate nutrition causes the feet to lose brightness. Their feet also get duller as they age, so Beatrice has her eyes on vibrant and virile younger males.

Looks are important, but they're not everything. As the saying goes: it's not about how blue your feet are; it's how you use them.

A male booby will strut his stuff in front of prospective mates, flashing his feet and subconsciously signalling that he's getting a surfeit of sardines in his diet. Then he'll crack out his secret move—

"sky pointing"—where he lifts his wings and tail, then raises his beak to the skies as if to thank the heavens for crafting such a fine specimen.

If Beatrice is suitably impressed, the two will get down to business. She'll then lay her eggs—usually a couple of them, but several days apart.

That gap is significant. If you've grown up as the youngest sibling then you'll know the pain it causes, but you should thank your lucky stars you're not a booby.

The first egg will hatch earlier and have an important head start on its sibling. If food is plentiful, they get along as well as can be expected. If there's a shortage, then all bets are off. The older sibling may peck its little brother/sister, steal its food, or even toss it out of the nest to its death.

You may be wondering where the parents are, as these squabbles escalate to full-blown siblicide. Well, the booby has a somewhat laissez-faire approach to parenthood. The parents are usually passive spectators, and they'll even reinforce the inequality by feeding the elder chick first.

It may seem callous, but it's a decent strategy to ensure at least one chick survives, rather than rationing food between two mouths. The younger sibling is an insurance policy, of sorts.

Siblicide is standard stuff for the animal kingdom, so the blue-footed booby is no worse a parent than many species. A better parent, in fact, than its relative—the masked booby.

Blue-footed boobies make steep-sided nests to deter big brother Boris from flinging tiny Tim to his death. Masked boobies don't take such precautions, and siblicide is far more common. (No wonder they grow up as fugitives, using masks to hide their identity.)

53

While we're on the subject of other species, let's investigate the indiscretions of Beatrice's ex.

First off, he wouldn't be a wandering albatross. While it's another large seabird, the wandering albatross is in a completely different family, and his cameo in this book was solely for a bad joke. Let's all wave goodbye to the hypothetical albatross as he soars majestically into the distance.

What about that search history? Don't worry; it's perfectly wholesome...unless you're Beatrice. (Perhaps those trust issues aren't unfounded.) Her ex could have been browsing saucy pictures of brown boobies, red-footed boobies, or even Peruvian boobies.

Fun fact: *Blue-footed boobies, although usually monogamous, sometimes practice bigamy. There have been cases of males sharing a nest with two females. (That's a lot of boobies!)*

Whether it's a traditional family of mum and dad, two chicks and a dogfish, or something more progressive, Beatrice needs to choose her next suitor wisely, because males play an important role in the rearing of young.

Parents share the incubating duties, so her new man will have to be sitting on a nice little nest egg. Males also provide for the chicks in the early stages, so he'd better put a diversified seafood portfolio together and can I stop with this analogy now please?

Let's hope romance is afoot for Beatrice.

Barbara, 19

Peace and love, dudes.

I'm a pansexual *Pan paniscus* who'll pounce on any pygmy primate with a pulse. (Although I prefer someone who comes from good breeding. Who was your mother?)

You can't be the jealous type, because I'm an extrovert and a social butterfly. Sharing is caring, and there's more than enough of me to go around.

The way to my heart is through my stomach, but I'd like someone who can stimulate my brain...as well as other parts of my anatomy.

Interests: I love a bit of pan piping, and I'm not talking about the instrument.

These promiscuous primates are a nice change of pace after some of the mating horrors we've seen so far. No love darts or penis fencing here.

Wait, there *is* penis fencing? We'll get to that later.

Bonobos (*Pan paniscus*) were historically called pygmy chimpanzees, though they're not *that* much smaller than their close relatives and are a distinct species. They're also the most mysterious—and perhaps least celebrated—members of the great ape family.

Why *is* that? I'm not sure. Perhaps they simply make the more puritanical among us a little uncomfortable. The gregarious great apes are one of the few species that has sex for pleasure. Not only that, but they'll also do the deed face-to-face and throw in the odd French kiss.

That's *more* humanlike than your (hypothetical) great uncle Albert, who did his duty with the lights off then prayed for forgiveness afterwards.

Not to mention, *they look a bit like us.* There's a reason Barbara's modesty has been preserved with a strategically placed leaf or two. This isn't that kind of book.*

What kind of book is it? Good question.

Bonobos don't just get freaky for fun. Sex also plays an important role in the social dynamics of their community. Bonobos mate to strengthen social bonds, to resolve conflicts, and even as a greeting.

Refreshingly, there's no stigma attached to same-sex relations. Females won't go more than a couple of hours without cementing their friendships by rapidly rubbing their (remarkably large, and often swollen) genitals together. It seems to be the bonobo equivalent of a catch-up over coffee.

Males aren't shy of showing their buddies they care, either. They'll rub their todgers together, occasionally even hanging upside down from a tree and "penis fencing." (Unlike flatworms, this doesn't end with one of them getting impaled. That's progress for you.) Rump rubbing reconciliations are another tool in their arse arsenal.

Despite their reputation as the hippies of the natural world, bonobo society isn't a carefree commune where everyone loves everyone else. There's a hierarchy in place, where females largely dominate the fellas. This is unusual in primates, and one of the major things that sets them apart from chimps.

There's still an alpha male, whose job it is to see off threats, but the high-ranking gals form their own gang and make most of the major decisions. A male gets his status from his mother—hence why Barbara's so concerned about good breeding.

This pecking order also impacts the *porking order*. That is, social hierarchy and sexual hierarchy go hand-in-hand. (Or hand-in-something-else.) Females can afford to turn down low-ranking males without worrying about being coerced.

Sometimes these loose rules go out of the window. For example, group sex is common after finding a new food source. A 1983 study in the San Diego Zoo found that:

"As soon as a caretaker approached the enclosure with food, the males would develop erections. Even before the food was thrown into the area, the bonobos would be inviting each other for sex."

Sounds like Friday night in a kebab shop in Liverpool. But I digress.

With the increased tribalism in human politics these days, it's almost inspirational to see that bonobos welcome strangers with open arms...and open legs. Male bonobos stay with the group they were born into, but females will often migrate to a new community during adolescence.

Unlike their territorial relatives the chimps, bonobos aren't so hostile to new faces. The young migrant females home in on high-ranking females, opting for the classic crotch rub technique to integrate into their new community.

Who says it's hard to make friends?

Conflicts are not unknown, but they rarely escalate to serious violence. Squabbles are usually followed by an apology which, naturally, takes the form of genital rubbing. Even separate communities have been witnessed "mingling" peacefully with one another.

Maybe humans could learn something from these progressive primates. It would certainly make political debates more interesting if we adopted bonobo social customs.

Shane, 12

Kelp! I need somebody.

I'm an impeccably groomed businessman with a successful shellfish empire, looking for No Swims Attached fun with a submissive saline sweetheart.

We can start out slow. Maybe hold hands and canoodle in the currents. But I have to warn you—I'm not like most otter guys. My tastes are very...*singular*.

What do you say? Let me *whisker* sweet nothings in your ear...then wrap you up in kelp and spank you like a disobedient weasel.

First Date: There are too many dive bars around here; let's go somewhere fancier.

Oh my. It seems everyone's favourite cute critter is hiding a few secrets in his underwater lair.

Sure, sea otters *look* adorable, but if we've learned anything, it's that appearances can be deceiving.

Perhaps being cute is part of the sea otter's sneaky strategy to cover up his crimes. Maybe we should be saying "aaaaaargh" instead of "awwww."

There's no doubt Shane's well put-together. Grooming is essential for sea otters, but it has a purpose that his human counterparts don't have to worry about—keeping him buoyant. Sea otters will clean their fur fastidiously, squeezing out water and blowing air into it.

Shane isn't monogamous and will mate with multiple females, but he does stick around and pair-bond with a single female while she's in estrus. After that, his parenting work is done, and he'll leave the pup in her capable paws.

The bond between a sea otter mother and her offspring is famously strong. Her dedication to her child—and the affection she showers it with—are almost enough to justify the otter's innocent and pure reputation.

Almost.

You've probably heard how sea otters will hold hands while they sleep to prevent them drifting apart. Romantic? Sure, but things get a lot raunchier than that.

Shane says he's not like most otter guys, but devious tendencies are all too common when it comes to this macho mustelid's mating strategies.

A favourite technique is to bite the female right on the nose and hold on while spinning around. This results in a lot of gals sporting some nifty scars. In extreme cases, these wounds have become infected and killed the female.

That's not all. Shane may even *hold her head underwater*. Female otters occasionally die from drowning when the male gets a bit carried away. (This is why safe words are important.)

Mercifully, I don't think spankings are part of the otter's bedroom repertoire, but they *will* wrap themselves in kelp to stay in one place and stop themselves floating out to sea. (Or, perhaps, to stop their victims getting away...)

You think that's it? You sweet summer child.

Sea otter males aren't discriminating, and they'll even target other species. Juvenile harbour seals are another unfortunate recipient of the lascivious lothario's "affections." Sadly, many are killed during the encounters. But that doesn't stop the otter from finishing...

In Shane's defence, consent isn't a concept many animals abide by in quite the same way as humans. "Forced copulation" has emerged as a mating strategy for numerous species, and we only seem to take notice when the perpetrator is considered cute.

That doesn't mean I'd let Shane within sight of my hypothetical daughter. He seems to be clenching that kelp a little too suggestively for my liking.

Agatha, 1

Just an Argonaut, waiting for my Jason to come aboard.

Ditch the other suckers, and shack up with this sexy cephalopod. I spent my gap year hitchhiking, but now I've built my own mobile home with a lovely ocean view.

I'm a shellfish lover, but I'm not a selfish lover. People have told me I ammonite-mare to date, but it's not like it'll cost you an arm and a leg...just an arm will do.

Even though I'm quite antisocial, I wish guys could be a cuddlefish after the deed is done, rather than always staying at arm's length. Premature evacuation isn't something you need to be embarrassed about, and I promise I probably won't eat you.

Favourite Joke: Any knock knock knock knock knock knock knock knock jokes.

Cephalopods are some of the planet's weirdest inhabitants, so it should come as no surprise that their mating habits have made it to this book.

Octopuses are intelligent, mysterious, and...*alien?*

Before you encourage me to take off my tin foil hat, hear me out. Scientific papers have suggested cephalopods may have originated away from earth and hitched a ride on a meteor.

Alright, no one's proven this, and other scientists have been quick to call theory wildly implausible. But I'm not seeing any better ideas.

So, Agatha might not be an alien. But even by cephalopod standards, the argonaut genus is full of octo oddballs.

Also known as the "paper nautilus," females craft an elaborate egg case that becomes their home *and* their mode of transport. Most octopuses hide their eggs away in safe places on the seafloor, but the argonauts are pelagic (open ocean) species, so Agatha improvises.

The paper nautilus moniker comes from said egg case, which slightly resembles the shell of another cephalopod—the nautilus. Nautiluses are the ancient (living) ancestors of young whippersnappers like squid and octopuses. They're related to the extinct ammonites, though their distinctive spiral chambered shell sets them apart.

Unlike their namesake, paper nautilus octopuses can leave the safety of their "shell," but spend most of their time tucked inside.

The word *nautilos* means "sailor," and Ancient Greeks (incorrectly) thought the octopus traversed the ocean in its own "boat," using two of its arms as sails. In fact, the female uses air bubbles to maintain buoyancy, and she moves by releasing trapped air.

She'd rather conserve energy, though. Being a pelagic species is tiring; Agatha's gap year was spent drifting along, "hitchhiking" a ride with passing objects...including jellyfish!

What about the octopus's other name?

"Argonaut" translates to "sailor of the Argo," which was the ship of the mythological Greek hero, Jason. In the legends, Jason and his crew (the Argonauts) sought the Golden Fleece. I think we're due an update of the legend, where Jason's a keen SCUBA diver who teams up with a band of octopuses to defeat Craig the King Crab.

Agatha's quest is the same as any animal's: to eat and mate. But it's the way she goes about the latter that's intriguing.

"Premature evacuation" isn't a typo, and it isn't anything for male octopuses to be ashamed of. It's how this genus mates. Males have a modified mating arm, called a hectocotylus, which is used to insert spermatophores into the female. (After he's bought her dinner, I hope.)

The argonaut is unusual among octopuses because the male's hectocotylus is *detachable*. When it's time for some hectic hectocotylus hanky panky, the male attaches himself to his lover, then sends his favourite (or possibly his *least favourite*) arm out on its own to finish the job.

By the time this disembodied arm has wriggled its way inside her mantle, he has already grabbed his coat and swum off, possibly after muttering: "this never happens to me, I swear."

In other words, Agatha is out of luck if she wants a guy to stick around for cuddles.

You can't blame them. Like many species with extreme sexual dimorphism, sexual cannibalism is common. Male argonauts likely developed the mating strategy to reduce their chances of being eaten by the much larger female.

Even if a male makes it out alive, his prospects aren't great. Males only mate once and seemingly die soon after, although octopus reproduction is still shrouded in mystery. (Only dead argonaut males have ever been found.)

The prospects for the female are decidedly better. She'll collect arms (and sperm) from several males, and she'll use them to fertilise her eggs over time.

Octopus mating, ladies and gents...just a bit of 'armless fun.

Colin, 3

Looking for my partner in crime.

I'm an internet personality and social media influencer looking for my next viral vid. You might remember me from such classics as "Pretending To Be A Sparrowhawk: Gone Violent" or "Mobbed By Warblers."

I've got something special planned for my prank YouTube channel. Let's just say it involves you, me, an egg, and someone else's nest.

I can't wait to have lots of little parasites (sorry, *chicks*).

Something You Should Know: I have a child from a previous relationship...and he's a murderer.

Ah, the sweet serenade of the cuckoo's call.
The sound of spring? Perhaps, if spring sounds like lies, misery, and broken homes.

Let's face it: Colin sounds completely obnoxious, which isn't unusual for most people who call themselves "influencers."

While cuckoos (to the best of my knowledge) don't have YouTube channels, some of the stunts they pull probably *would* go viral.

The cuckoo is what's known as a brood parasite. (Females lay their eggs in other species' nests.) While she's not going to win Mother of the Year, it's a solid strategy for passing on her genes without investing any time or energy raising a chick.

That responsibility falls on the poor sucker whose nest the egg is laid in—*unless* she realises the ol' switcheroo has taken place. Making sure the host mother doesn't suspect treachery has led to a sort of arms race between cuckoos and hosts.

The female cuckoo will swoop in, push an egg out of the host's nest, then replace it with at least one of her own (which, although larger, closely resembles the host's eggs). Even within a single cuckoo species (in Colin's case, the common cuckoo), different individuals will target different host species.

Some female common cuckoos have evolved to hoodwink reed warblers, while others target robins, and so on. Despite being the same species, common cuckoos' eggs differ depending on their host.

As for Colin? He can breed with any female of his species, and he'll also be playing his part in this operation, even if it just involves standing there looking intimidating.

The videos he referenced in his profile are relevant. Cuckoos have a lot of sneaky strategies in their repertoire for infiltrating an unwitting bird's nest with their own eggs.

Male common cuckoos actually *do* pretend to be sparrowhawks. Not consciously, but they've evolved similar plumage to the bird of prey, which helps to scare off host parents and gives the female cuckoo a window of opportunity to carry out *Mission: Immoral*. She only needs seconds to replace one of the eggs with her own.

If the hosts see her, or realise the male isn't a sparrowhawk, they'll often respond aggressively and try to chase off or "mob" the larger cuckoos.

Even if the plan goes smoothly, there's no guarantee of success. As cuckoos evolve better egg mimicry, the hosts are also evolving to detect the egg imposters better. (After all, their species wouldn't last long if they *always* raised cuckoos instead of their own young.)

Still, much of the time the host won't suspect a thing. If the mother accepts the egg, she and her partner will end up raising a huge chick that looks nothing like either "parent."

The cuckoo chick doesn't like sharing food, so it will often evict its adopted siblings (whether they've hatched or not) by pushing them to their death.

Like mother, like child...

This fat, fluffy felon will swiftly outgrow mummy and daddy, which no doubt leads to some awkward conversations between Mr and Mrs Reed Warbler as to the parentage of their child.

At least that's what *should* happen.

In reality, once Colin Jr has hatched, the host parents don't suspect a thing. Perhaps it's because he's a talented mimic, and he makes a begging call that sounds like the host chicks he heartlessly murdered.

I prefer to believe the parents are in denial when, after working themselves to the bone feeding his cavernous mouth, he tells them he wants to be an influencer when he grows up.

Manny, 8 hours

Hurry up.

I live each day like it's my last...

Because it is.

#YOLOD

Favourite Season: Afternoon.
Favourite Minute: 26.

I n an ideal world, we'd take our love lives slowly, getting to know prospective partners as romance begins to take hold and grow like a parasitic fungus. (Err, I'm not too good at similes.)

Unfortunately, this isn't a viable mating strategy for mayflies, because they don't have the luxury of *time*.

Mayflies are the animal kingdom's poster child for short-lived species, and it's true that they don't hang around. The well-known figure is 24 hours, though there are over 3000 species, with adult lifespans ranging from a few days to a few *minutes*.

I'm sure you've heard plenty of artsy metaphors about the ephemeral nature of this insect's existence, and how it packs so much into a few precious hours. There's something wistfully romantic about it, though I'm not sure the mayfly shares our sentiments.

"Carpe diem," we might say, if we're feeling particularly unoriginal. *Seize the day*, we think to ourselves, as we share some motivational quotes on Facebook.

"Carp! Aaaargh!" the mayfly says. *Seize the prey*, the fish thinks to itself, as it shares some motivational quotes on Plaicebook.*

You can't hate me more than I hate myself.

See? We're all intertwined. All part of life's rich tapestry. All trying to find meaning; trying to survive; trying to find love...Oh yeah, *that's* what we're supposed to be talking about. This topic just brings out my inner poet.

As I was getting at earlier, Manny needs to pull his finger out and find a gal sharpish, because females live even shorter lives than males.

Like many insect species, young mayflies look completely different to adults. Uniquely among insects, though, mayflies have a winged, pre-adult stage between their immature (nymph) and adult forms, called a "subimago." This moult may last a day, or even a few minutes for the shorter-lived species.

Fun fact: *Ironically, the mayfly order Ephemeroptera is one of the most ancient of all living insects.*

If the subimago hasn't been picked off by predators, it swiftly moults again into its final, adult (imago) form, raising the question: why didn't it just do that in the first place?

Manny's favourite saying: YOLOD (You Only Live One Day), was optimistic. The adult female *Dolania americana*, for example, finds a mate, deposits her eggs into the water and dies, all within the space of *five minutes*. The males might last half an hour, patrolling the water for females before dropping with exhaustion and drowning.

There's a caveat. While the *adult* (or pre-adult) mayfly isn't winning any awards for longevity, they spend the majority of their lives underwater as nymphs. These immature insects can live in their streams or lakes for a couple of years before undergoing metamorphosis.

So, I suppose the mayfly *actually* teaches us to spend most of our lives as ugly babies lurking in a lake, before dying as soon as things finally get interesting.

Clownfish

Claire, 5

Am I a joke to you?

I'm a self-made woman (last week I was called Clarence) with my own place, looking for a worthy suitor to step up. You'll have to be immune to stings, but not to my charms.

I already have a symbiotic relationship with someone, but I'm hoping you can be friends rather than anemones.

My work? I know it's a cliché, but I'm a comedian. And yes, women *can* be funny, thank you very much. I sold out the Coral Club with my autobiographical special: *Finding Myself*.

Perfect Date: Stay in and snuggle. (I'm kind of agoraphobic.)

73

I n a revelation that should surprise precisely nobody, it turns out *Finding Nemo* would be a whole lot weirder if it were true to life.

Let's focus on what it got *right*, first.

Clownfish, which comprise over 30 species, are known for their symbiotic relationship with anemones. They're either immune to the venom of the stinging tentacles, or they're protected by a mucus coating.

The relationship is mutually beneficial. The anemone gives the fish shelter and protection from predators, as well as a safe place to breed and lay eggs. Clownfish tend to keep their friends close and their anemones closer, as they're much more vulnerable to predators in open water.

As for the anemone? It doesn't mind having the fish around because they prey on the anemone's parasites...*and poop on it.* That may not sound like a good thing, but clownfish excrement contains nutrients that the anemone craves.

Nemo and his father Marlin (as well as Claire) are ocellaris clownfish, which can live in three different anemone species. These clownfish *can* make great fathers, at least before the eggs hatch. Males do most of the "egg sitting," carefully fanning the eggs with their flippers to give their offspring more oxygen.

Once the young hatch, they're on their own. They drift out into the open ocean, tiny and vulnerable, before the survivors seek their own anemone.

So far, so good. We can forgive children's movies for not focusing on the poopy particulars of symbiosis or spreading the moral message that kids should be kicked out at birth. But things are only about to get weirder.

Clownfish society is not all sunshine and rainbows. There's a strict dominance hierarchy within the anemone. The largest, most aggressive female lords it over her subjects, and she'll only mate with one other male. The rest of the group, who are juvenile males, don't have much fun. They're submissive, they never get laid, and they're bullied by the dominant breeding pair.

What's a juvenile male to do? Well, he (yes, always *he*) could try his luck moving to a different anemone, or he could just suck it up and wait for the dominant female to die. When that happens, every clownfish in the colony moves up one rank in the hierarchy.

Ah, but there's still the small issue that *the only female just died.*

Well, that's *not* actually an issue, because clownfish are sequential hermaphrodites. All of them develop into males first, then only some will mature into females.

That means the dominant male will just *become* the dominant female if the old one dies, leaving a vacancy for a new dominant male from the ranks of the juveniles.

Claire wasn't lying when she said she was a *self-made* woman.

So, a more realistic *Finding Nemo* would see Marlin become Nemo's mum, then shack up with her new beau, which—in the absence of any other males—would be, err...*Nemo himself.*

I can see why they went in a different direction.

Bowerbird

Brian, 7

I'm lonely and I'm blue.

Do you want a man who's good with his hands? Well, that's a bit unrealistic. I'm good with my beak *and* my feet, though.

I heard you like artistic avians, so I built this shrine of seduction from scratch. It just needs a few licks of paint and it'll be the perfect birdy boudoir.

Let me take you under my wing and shower you with bling. I love to treat my woman with trinkets, just don't ask where I got them.

Fine. They, err, fell off the back of a truck.

Favourite Colour: BLUE!

Looks like we're going back down under to investigate an Aussie artist-cum-architect.

While his fellow countryman, the antechinus, lacks subtlety in his seduction, the male satin bowerbird isn't afraid to show off his creative side. He *has to*, because competition is fierce, and he's not much of a fighter.

This raunchy rainforest Rodin sculpts his shrine of seduction out in the sticks, made from...sticks. The structure, known as a bower, is then decorated with assorted colourful trinkets—some natural, and some "borrowed" from the bird's two-legged neighbours.

When he says it needs a lick of paint, he's not lying. Thrifty satin bowerbirds will paint the walls of the bower with chewed up vegetable matter and saliva, saving a fortune in the process.

The purpose? To impress the ladies. Building and decorating a bower is an important part of the bird's elaborate courtship ritual.

A bower may be festooned with feathers, flowers and berries, or even plastic straws and bottle caps...whatever comes to beak.

Different species have different preferences. The rather macabre great bowerbird decorates his bower with bones, possibly to attract goth birds going through their youthful rebellion phase.

Brian—and other satin bowerbirds—have a thing for the colour blue. Males will scour their surroundings, and they're not above stealing from their rivals.

Female satin bowerbirds get the cushier job, at least to start with. They strut around the forest, inspecting potential suitors' bowers like judges on a reality TV show. They'll make several visits, slowly whittling down the contestants.

What are they looking for? It's hard to tell. The bower isn't a nest; its value is purely aesthetic. And once they've mated, the male will leave the chick-rearing duties entirely to the female. That means his genes are crucial.

Perhaps she's hoping she strikes the jackpot. He may be a penniless artist now, but with a stroke of luck—and the right publicist—he could have a successful career displaying his contemporary artwork at galleries around the world.

It's not just the bower that matters. Males also need to show off their fancy footwork and display that striking blue-black plumage. Brian could have all the charisma in the world, but if he has a dad bod and two left feet then he doesn't stand a chance.

Younger females are more swayed by the appearance of the bower, while more mature gals place more value on the male's dancing skills. They know it's a tough job market for liberal arts graduates, but a dancer will always put fruit on the table.

Looks like Brian's got his work cut out. Females can choose to be picky; they'll be the ones raising the chicks, after all.

Marty, 5
Quack quack, watch your back.

I'm not here to take part; I'm here to take over.

Some call me invasive... I prefer the term *cosmopolitan*. I love to see the world. Dabble in different cultures. Ponder the pond life. Break bread with the natives.

My type? If you've got webbed feet and wings then you fit the bill, but I like a feisty firequacker who plays hard to get.

Do I come on too strong sometimes? Maybe, but you know what they say—no harm, no fowl.

Hidden Talent: Let me show you how I open a bottle of wine.

T he humble duck.
    A figure of fun, to many people. Waddling along mer-
    rily. Paddling lazily on a pond. Gobbling up that bread you
shouldn't be feeding it. (It's bad for them.)

To that, I say: *you don't know ducks*. Especially mallards. These crafty quackers are the living embodiment of fowl play.

If you'd like to live in blissful ignorance, then feel free to turn the page, where there are more wholesome stories of romance and...err, let's be honest—the other profiles aren't much better.

The mallard is a widespread species in the wild and the main ancestor of many domestic ducks. It's basically a pond-dwelling Genghis Khan, taking over territory and mating with everything in sight.

"Genetic pollution" sounds dangerously like something an unhinged dictator might warn against. Conservationists are rightly concerned about the dangers, though. The adaptable mallards are considered an invasive species in many areas they've been introduced, outcompeting native species or polluting their genes via interspecies intercourse. This can create fertile hybrids, potentially causing extinctions of the original species.

Maybe you think the mallard should be free to mate with whomever it desires? Don't come crying to me when one day, you wake up and *everything is a mallard*.

When it comes to mating, this dastardly dabbling duck likes to dabble in all sorts of shenanigans. We've already explored "forced copulation" with Shane the sea otter, but mallards take it to the next level.

It makes evolutionary sense for females to be picky. They want the father of their offspring to bring some good genes to the table. In mallard terms, that means handsome plumage and an impressive display.

Unfortunately, this results in a lot of scorned male suitors, who attempt to get what they want by force. In fact, males' preference for the rough stuff has led to—you guessed it—an evolutionary arms race.

But it's not their arms that are evolving...

Yeah, we're talking about weird genitals again.

Yale ornithology professor Richard O. Prum dedicated a chapter to the dark side of duck mating in his book, *The Evolution of Beauty*. He writes about "pervasively common" forced copulation, including "violent, ugly, dangerous and even deadly" gang rapes.

This sounds pretty awful for females, and it doesn't get better when you find out what kind of equipment drakes are packing.

Ninety seven percent of bird species don't even *have* a penis, but many male ducks are well-endowed with long, spiral phalluses that resemble corkscrews.

"Like a selection of sex toys from a vending machine in a strange alien bar," Prum writes, "duck penises come in ribbed, ridged and even toothy varieties."

Females have found a way to fight back, though. Not by *preventing* the forced copulation, unfortunately, but by preventing it from being successful.

Their own reproductive anatomy has evolved to make it hard for any but the biologically "fittest" males to copulate. To counter the explosive corkscrew cocks of doom, many female ducks boast anti-corkscrew-shaped vaginal tracts.

"Male ducks had evolved penises that would enable them to force their way into an unwilling female's vagina, and the females in turn had evolved a new way—an anatomical mechanism—to counter the action of the explosive corkscrew erections of male ducks and prevent the males from fertilizing their eggs by force," Prum writes.

It seems counter-intuitive that females' anatomy would evolve to make mating *harder*. But if the female wants to mate, she can make it easier for her partner by relaxing the walls of her genital tract. Conversely, if the drake hasn't got consent, she'll do all she can to prevent him getting her pregnant.

It must work, because forced copulation is extremely common, yet only a small proportion of duck offspring are born from these encounters.

Females once again seem to get the short end of the stick, but male mallards aren't immune to punishment. Occasionally these randy drakes will chase and attempt to mate with other males. Or even, erm, *ex males*.

One noble scientist wrote a paper entitled: "The first case of homosexual necrophilia in the mallard." The titular mallard drake chased a fellow male, who flew off, promptly collided with a window, and died.

"Ah well, better luck next time," the surviving drake *didn't* say, because he proceeded to have sex with the dead duck's corpse for the next 75 minutes, until the author of the paper shooed him away.

Maybe I should delete Marty's profile. There's enough suffering in the world.

Georg...ina, 3

Let's make hisssstory.

It's been a cold winter. Is there a saucy serpent who can raise this cold-blooded gal's body temperature and make me feel alive again?

Scratch that; the more the merrier. Let's make this a group date, but it's up to you to make the first move. I'm used to being the belle of the (mating) ball, so you'll have to chase me.

You won't be able to resist my enticing aroma. You know what they say: pheromones lead to pheromoans.

Let the games begin.

Favourite Movie: *Snakes on a Plane.*

I'm sure many of us have had our share of dishonesty-related mishaps on dating apps. Whether it's a decade-old profile picture or the cannibalistic cult they failed to mention (don't ask), the dating world can be a treacherous place. But humans have got nothing on garter snakes, who take trickery to a whole new level.

Garter snake group sex is a well-known phenomenon, and the large "mating balls" they form are the stuff of legend. Dozens of males flock to a single female, who soon finds herself at the bottom of a writhing ball of scaly suitors.

Where does the trickery come in? Well, sometimes those females are not females at all, but male imposters.

Why would males pretend to be females, then? After all, being at the receiving end of a garter gang bang doesn't sound particularly pleasant.

All shall be revealed.

There are many different subspecies of the common garter snake; perhaps the most-studied is the red-sided garter snake. Manitoba, Canada, has become an unlikely tourist destination, and for a few weeks each spring the Narcisse Snake Dens are packed with binocular-wielding perverts (possibly) straining to get a glimpse of some sweet snake action.

Actually, I doubt binoculars are necessary, because these exhibitionists (the snakes, I mean) gather in their thousands for these outdoor orgies, and they don't care who's watching.

84

The garter snakes emerge from brumation (a kind of mini hibernation during cold weather), but it's the males who wake up first and warm up a bit for the main event. They've got one thing on their minds, and it's not finding someone who can tailor them a bespoke blazer.

Once the females emerge, the horny hissers converge on them, forming a writhing mass of raunchy reptiles. The females give off irresistible pheromones that tell the males they're single and ready to mingle.

Some mischievous males have learned to take advantage of this and can produce pheromones that mimic the females' own seductive scents.

That's right—Georg...ina is actually a George.

But why does he want to be at the business end of a snake orgy?

It's not just because he's a kinky weirdo. Being at the bottom of all those bodies means he's getting quite nice and toasty, and for a cold-blooded snake, warming up is crucial.

It's also a lot safer at the bottom of the pile. Predators converge on the mating balls, looking for an easy meal of distracted snakes on the lookout for some tail.

Once his rivals have wasted their energy pursuing him, or been picked off by predators, George will have a decent chance of finding a real female of his own to mate with.

Unless, of course, he's hoodwinked by another male doing exactly the same thing.

Survival of the fittest? More like survival of the sneakiest.

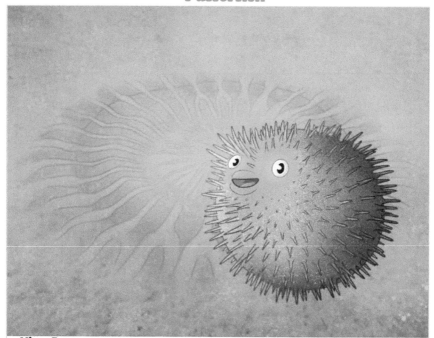

Hiro, 5
Konnichiwa.

I have given in to my base desires, and seek carnal pleasures amongst the sands of the sea. There is a certain inevitability to it; toxicity is within the very core of my being.

The pattern is already hidden on the ocean floor. I merely remove what obfuscates it, just as I remove worries from my mind. The peaks and valleys of the circle mirror those of our own lives. Like all earthly beauty, it is transient. Fleeting. Washed away with the currents.

A shifting sediment of seduction. A momentary monument to our own impermanence.

I cannot bend the sands to my will, nor tell the waves to leave my efforts untouched. Yet I have seen it. And you will have seen it. And that is enough.

And then we're going to bang.

You're probably wondering what the heck just happened, and I don't blame you.

Do the words "artistic pufferfish" help at all?

No? Fair enough. I didn't think they would.

Well, despite the slight inconvenience of not having hands, this little Zen master manages to create beautiful artwork to woo his lady love.

I say Zen, because the aquatic artist in question is found in Japanese waters. And his artwork does seem to fit the simplistic, natural aesthetic, or wabi-sabi, where the beauty of imperfection and impermanence is celebrated.

Hiro is a *Torquigener albomaculosus*, or white-spotted pufferfish, a species which was only discovered several years ago. But he's not like other pufferfish; he's quite extraordinary because of the huge, geometric circular nests he crafts on the ocean floor.

Despite being a modestly sized chap, at only 10 cm or so in length, his nests can be two metres in diameter and take over a week to create. These "crop circles" were a bit of a mystery to science, until the little guy was caught in the act.

He flaps his fins furiously, slowly disrupting the sand as he laboriously crafts the peaks and valleys of his pièce de résistance. This causes fine sediment to drift towards the centre of the circle. He'll then decorate the outer circle with shell fragments. Finally, he'll make irregular patterns with the sediment at the circle's centre.

So, *why* does he go to the trouble of constructing this masterpiece? It's especially baffling because, as he said so poetically, the art is transient and washes away with the currents.

Scientists have investigated, and they know it's something to do with attracting females, but they're not sure *what* females are looking for when they inspect their prospective mate's artwork. Whether it's the size, symmetry, decoration, or the circles are just a mechanism for moving the fine sediment to the centre, the jury's out.

After scratching her chin and saying: "somehow by saying so little, it says so much," the female will lay her eggs in the centre of the circle, then leave. The male will stay, first to fertilise the eggs, then to guard them for six days.

After the eggs hatch, the male moves onto a new site and starts again from scratch.

So Zen.

Things could get a little more stressful if he sees a predator, in which case he might inflate himself. (There's a reason he's called a pufferfish, after all.)

A fat, spiny ball isn't a particularly appetising prospect, but that's not his only defence. Most pufferfish are highly toxic, and their internal organs produce a potent neurotoxin called tetrodotoxin.

I guess he wasn't just being pretentious about toxicity after all. But I probably was with my interpretation of Zen artwork.

Ah well.

Echidna

Ernie, 4

Fresh out of hibernation, and looking for some titillation.

Roll up, roll up, and get an eyeful of your dream monotreme. (Actually, don't roll up; it'd make things tricky.)

I'm on the lookout for a saucy Sheila to root around the underbrush, if you catch my drift. I may be short-beaked, but I'm long where it counts. And you know what they say: "two heads are better than one, but four heads are better than two."

If you let me join your love train I'll show you I can outlast those other spiny suckers.

Favourite Song: "Love Train" by the O'Jays.

D id you think you'd heard enough about weird animal penises? (That's rhetorical.)

When it comes to the Dubious Genital Olympics, the echidna is a worthy contender for gold, and I'd be remiss if I didn't mention its multi-pronged member. That's right—the male echidna has a four-headed phallus.

The four species of echidna, along with fellow Aussie oddity the platypus, are the only egg-laying mammals in the world, otherwise known as monotremes. Ernie is a short-beaked echidna, the only species native to Australia. (His long-beaked relatives live in New Guinea.)

The name "monotreme" translates to "single opening," referring to the cloaca, a single duct responsible for urination, defecation, and reproduction.

Ernie is equipped with an impressively odd *implement* which is normally retracted inside his cloaca, but emerges in all its glory when it's time for some baby making. (Or puggle producing, to be more species-specific.) After mating, a single egg forms in the female's pouch, and a baby echidna (called a puggle) cracks its way out.

But that's the aftermath; I'm sure you want to know more about the deed itself.

Echidnas quickly transition from hibernation to procreation. At the start of the mating season males and females turn their cloacas inside out and wipe them on the ground, creating a musky odour that tantalises the opposite sex's olfactory system.

So far, so relatable, but what happens next is a little more unusual.

Sometimes, several males pursue a single female, quite literally forming a queue, with the youngest, smallest individuals bringing up the rear. This weird courtship ritual can last a few weeks, with some males dropping off and giving up the pursuit, while others join the love train.

Males sometimes fight for the right to mate, but it's still up to the female, who has the handy option of rolling into a ball if she's not impressed by her suitor.

Once she's accepted a mate, they'll dig a small crater and lay side-by-side. The male then shows off his favourite party trick, whipping out his spiny, four-headed phallus. He doesn't actually *use* all four heads at once, since the female only has a two-branched reproductive tract.

To get around this issue, the male alternates between sides, with two heads swelling up while the other two shut down (and vice versa, the next time he mates).

It's impressive that we even know this, because echidna sex is notoriously difficult to witness. In fact, prior to 2007, no one had seen an echidna ejaculate.*

First man on the moon? Pah. Neil Armstrong has got nothing on Steve Johnston, the first man to see an echidna bust a nut. "One small step for man, one giant, spiny erection for monotremekind" is what he (should have) said.

What's equally odd is the method used to try and extract semen from the echidna. The scientists *electrocuted a male's penis*, causing it to swell up into "a four-headed monster that wouldn't fit the female reproductive tract."

Eventually, scientists managed to work out that males only used two heads at a time, so let's hope captive echidnas will now be treated in ways that don't breach the terms of the Geneva Convention.

What's that, he requested that scientists *keep doing it?*

Different strokes for different folks.

Percy, 6

Single and ready to tinkle.

Something potent is wafting on the wind, and it's time to find my spiny soulmate.

You can call me Shakespeare, because I'm handy with a quill. *Shall I compare thee to a Summer's day? Thou art hot.*

I wrote that myself, but I have another medium in mind for my latest masterpiece. Let's just say urine for a treat.

My phallic paintbrush is in hand, and you, my dear, are my canvas. Once you're suitably splattered, we can (carefully) commence the conception.

You might feel a little prick.

First Date: How about a romantic shower?

Arboreal assaults, golden showers, stinky scents, vaginal plugs, and a sprinkling of stabbings—no one ever said porcupine sex was easy. (Nor should you, at least not in polite conversation. Probably why I never got invited back to that dinner party.)

Luckily for you, this isn't a dinner party, so let's get to the bottom of those bad puns and investigate the porcupine's sexual proclivities. I'm focusing on the New World porcupines like Percy, who's a North American porcupine.

We start off, as we so often do, with an irresistible musky mucus the female secretes from her vagina. This mixes with her urine to form a potent concoction, which signals that she's receptive to some raunchy rodent action.

North American porcupines are mostly solitary until the breeding season, but when Percy gets a waft of that flirty fragrance, he hightails it to the female's tree and starts climbing.

He'll stay on a lower branch for a while, biding his time. Perhaps he's working up the courage for what comes next; perhaps he's working on some more poetry. (It certainly needs work.) He can't take too long, because she's only fertile once a year, for a matter of hours.

If other males have the same idea, he might need to fight for his right to mate. The loser will limp off with a few quills to the face for his troubles; the winner will unzip his flies and—there's no polite way to say it—pee all over his prickly partner with a high-velocity jet of urine. He'll rear up on his hind legs, aiming his stream at the female up in her branch and firing off several salvos.

Projectile pee soaking isn't generally advised as a seduction technique in the human dating world, but apparently it does the trick on female porcupines.* A chemical reaction occurs, causing her to enter estrus. She'll clamber down from the tree, and the two will mate on the ground.

*It's not always successful. If she doesn't like the cut of her suitor's jib, she'll subtly convey this to him by screaming, biting him, and/or running away.

How do porcupines have sex? *Carefully.*

Luckily for the male, the female is considerate enough to flatten her quills and lift her tail, allowing him to gingerly do the deed. *Unluckily* for the female, the quills aren't the only sharp pieces of the male's anatomy. His penis itself is covered in hundreds of tiny spines.

Maybe that explains all the screaming.

Once the female has had enough, she'll head back to her tree. The male will go off to find more victims. The female forms a "vaginal plug" that increases the chances she'll get pregnant, while preventing other males from trying their luck.

Almost seven months later, a little porcupette will be born. (Yes, they're called that, and yes, it's one of the few adorable parts of this tale.) If you're wondering, the spikes don't harden until *after* the mother has given birth.

After being splattered with urine and impaled with a prickly penis, I think we can all celebrate a small win there for Mrs Porcupine.

With Gary Gorillason's warning still ringing in the author's ears, Alex disregarded his friend's sage advice and rolled out his animal dating app to worldwide fanfare.

Our team tracked down some of the animals who starred in the book, to see how their quest for love had progressed after several months on the online dating scene.

It wasn't pretty.

Adele amassed an impressive pebble collection from her new boyfriend, Arthur. Having got what she wanted, she ran off with an extra from March of the Penguins. Arthur was court-ordered to make chick support payments of regurgitated fish guts every month.

Marty is wanted in 16 countries for crimes against waterfowl. He was last seen groping a gadwall in rural Louisiana. Authorities have warned that he's armed and dangerous, but he may retreat if you pelt him with stale bread.

Shane found a timid significant otter, and the two had a tempestuous romance that was later made into a commercially successful (but critically panned) series of erotic novels. To everyone's great relief, Shane was bitten in half by an orca on the first day of filming the movie trilogy.

Kayleigh grew tired of the dunces she found on the local dating scene and started looking into her convict ex-boyfriend's case, saying she'd uncovered "The Great Fingerprint Conspiracy." The Australian Government categorically denies that koalas are being framed for human crimes. Meanwhile, Kayleigh has disappeared. According to local reports, she was plucked from her tree (in rural NSW) by a great white shark.

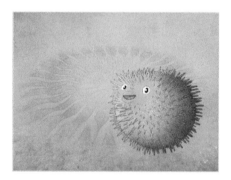

Hiro seduced several subaquatic sweethearts with his breathtaking sand sculptures, but he strayed from the path of wabi-sabi after starring as a judge on Ocean's Got Talent, where he rated other creatures' creations. He's currently the mascot for Fugu Flakes Breakfast Cereal in Japan.

Alex Cooper is a freelance writer and author of between 1.5 and 2.5 books, who currently lives with his inner demons in rural Norfolk, England.

When he's not writing about weird animal penises, he enjoys long, romantic walks to the fridge, procrastinating, and signing off *About the Author* sections with completely irrelevant quotations.

"Ski-bi dibby dib yo da dub dub, yo da dub dub. Ski-bi dibby dib yo da dub dub."

Scatman John

Also By Alex Cooper

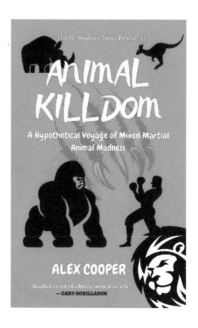

"If you ever wondered which partner a seemingly outmatched monotreme would tag-in to best a human opponent, then this is your source."

Dave Todd

"Well written, easy to read and a bit mad... in a good way."

Malky McEwan

"Brilliantly written, funny as hell, and just plain entertaining. And at the end of the day, genuinely educational."

T. Divens

CPSIA information can be obtained
at www.ICGtesting.com
Printed in the USA
BVHW091443111220
595480BV00009B/129